Table of Contents

INTRODUCTION 4

CHAPTER ONE 6

CLIMATE CHANGE HISTORY 6

    Early Inklings That Humans Can Alter Global Climate 7

    The Greenhouse Effect 7

    Greenhouse Gases 9

    Welcoming a Warmer Earth 11

    Keeling Curve 12

    When did scientists first warn humanity about climate change? 13

    Why the 1950s? 16

    1970s Scare: A Cooling Earth 22

    Global Warming Gets Real in 1988 23

    IPCC 23

    Kyoto Protocol: United States In, Then Out 24

    An Inconvenient Truth 25

Paris Climate Agreement: the United States In, Then Out 26

Greta Thunberg and Climate Strikes 28

CHAPTER TWO 29

CLIMATE CHANGE AND OUR HEALTH 29

What is climate change, and how is it affecting Earth? 29

How do scientists know climate change is real? 31

How is the climate changing? 33

What are the impacts of climate change? 35

Can we reverse climate change? 38

8 ominous climate milestones reached in 2021 42

Paris Agreement warming targets surpassed 42

Record-breaking heat in 2020 43

Faster sea level rise 44

Gulf Stream slowdown 45

Human influence 'unequivocal' 46

Carbon factory rainforests 47

'Last Ice Area' melting away 48

Earthshine gets darker 49

Humanity faces a' grave and mounting threat' of climate change 51

Human Health Impacts of Climate Change 54

Still not too late 56

Can we adapt? 58

Climate-induced Natural Disaster 59

Becoming Resilient Together 62

Expert Recommendation About Climate Crisis 63

Five steps to restrain emissions right now 66

CONCLUSION 69

INTRODUCTION

The climate crisis is accelerating at a dizzying pace and is the single planetary emergency that will determine our fate. Earth's climate has changed throughout history. Just in the last 800,000 years, there have been eight cycles of ice ages and warmer periods, with the end of the last ice age about 11,700 years ago marking the beginning of the modern climate era — and human civilization. Most of these climate changes are attributed to very small variations in Earth's orbit that change amount of solar energy our planet receives.

The current warming trend is different because it is the result of human activities since the mid-1800s, and is proceeding at a rate not seen over many recent millennia. It is undeniable that human activities have produced the atmospheric gases that have trapped more of the Sun's energy in the Earth system. This extra energy has warmed the atmosphere, ocean, and land, and widespread and rapid changes in the atmosphere, ocean, cryosphere, and biosphere have occurred.

Climate change, together with other natural and human-made health stressors, influences human health and disease in numerous ways. Some existing health threats will intensify and new health threats will emerge. Not everyone is equally at risk. Important considerations include age, economic resources, and location.

The effects of the climate crisis are happening right now. From natural disasters to supply chain shortages. The health effects of these disruptions include increased respiratory and cardiovascular disease, injuries and premature deaths related to extreme weather events, changes in the prevalence and geographical distribution of food- and water-borne illnesses and other infectious diseases, and threats to mental health.

More essentially, in this handbook, you will find practical ways- as provided by experts, on how to analytically handle the existing effects and coming ones of climate change on us and our environment.

CHAPTER ONE
CLIMATE CHANGE HISTORY

Climate change is the long-term alteration in Earth's climate and weather patterns. It took nearly a century of research and data to convince the vast majority of the scientific community that human activity could alter the climate of our entire planet.

In the 1800s, experiments suggesting that human-produced carbon dioxide (CO_2) and other gases could collect in the atmosphere and insulate Earth were met with more curiosity than concern.

By the late 1950s, CO_2 readings would offer some of the first data to corroborate the global warming theory. Eventually, an abundance of data, along with climate modeling would show not only that global warming was real, but that it also presented a host of dire consequences.

Early Inklings That Humans Can Alter Global Climate

Dating back to the ancient Greeks, many people had proposed that humans could change temperatures and influence rainfall by chopping down trees, plowing fields, or irrigating a desert.

One theory of climate effects, widely believed until the Dust Bowl of the 1930s, held that "rain follows the plow," the now-discredited idea that tilling soil and other agricultural practices would result in increased rainfall.

Accurate or not, those perceived climate effects were merely local. The idea that humans could somehow alter climate on a global scale would seem far-fetched for centuries.

The Greenhouse Effect

In the 1820s, French mathematician and physicist Joseph Fourier proposed that energy reaching the planet as sunlight must be balanced by energy returning to space since heated surfaces emit radiation. But some of that energy, he reasoned, must be held within the atmosphere and not return to space, keeping Earth warm.

He proposed that Earth's thin covering of air—its atmosphere—acts the way a glass greenhouse would. Energy enters through the glass walls but is then trapped inside, much like a warm greenhouse.

This theory was further explored by the work of Eunice Newton Foote in the 1850s. Foote's experiments using glass cylinders demonstrated that the sun's heating effect was greater in moist air than in dry air.

She detected the highest degree of heating occurred in a cylinder containing carbon dioxide. Her work would foreshadow the work of Irish scientist John Tyndall who also zeroed in on what kinds of gases played the biggest role in absorbing heat.

Experts have since understood that the greenhouse analogy was an oversimplification since outgoing infrared radiation is not exactly trapped by Earth's atmosphere but absorbed. The more greenhouse gases there are, the more energy is kept within Earth's atmosphere.

Greenhouse Gases

But the so-called greenhouse effect analogy stuck and some 50 years later, the work of Eunice Newton Foote offered further insight into how heat could be absorbed into Earth's atmosphere.

In the 1850s. Foote's experiments using glass cylinders demonstrated that the sun's heating effect was greater in moist air than in dry air. And she detected the highest degree of heating occurred in a cylinder containing carbon dioxide.

Although Foote, an amateur scientist, was never recognized in her lifetime, her work foreshadowed the findings of Irish scientist John Tyndall. Tyndall also explored exactly what kinds of gases were most likely to play a role in absorbing sunlight.

Tyndall's laboratory tests in the 1860s showed that coal gas (containing CO_2, methane, and volatile hydrocarbons) was especially effective at absorbing energy. He eventually demonstrated that CO_2 alone acted like sponge in the way it could absorb multiple wavelengths of sunlight.

By 1895, Swedish chemist Svante Arrhenius became curious about how decreasing levels of CO_2 in the atmosphere might cool Earth. To explain past ice ages, he wondered if a decrease in volcanic activity might lower global CO_2 levels.

His calculations showed that if CO_2 levels were halved, global temperatures could decrease by about 5 degrees Celsius (9 degrees Fahrenheit).

Next, Arrhenius wondered if the reverse were true. Arrhenius returned to his calculations, this time investigating what would happen if CO_2 levels were doubled. The possibility seemed remote at the time, but his results suggested that global temperatures would increase by the same amount—5 degrees C or 9 degrees F. Decades later, modern climate modeling has confirmed that Arrhenius' numbers were not far off the mark.

Welcoming a Warmer Earth

Back in the 1890s, however, the concept of warming the planet was remote and even welcomed. Arrehenius wrote, "By the influence of the increasing percentage of carbonic acid (CO_2) in the atmosphere, we may hope to enjoy ages with more equable and better climates, especially in the colder regions of the earth."

By the 1930s, at least one scientist would start to claim that carbon emissions might already be having a warming effect. British engineer Guy Stewart Callendar noted that the United States and North Atlantic region had warmed significantly on the heels of the Industrial Revolution.

Callendar's calculations suggested that a doubling of CO_2 in Earth's atmosphere could warm Earth by 2 degrees C (3.6 degrees F). He would continue to argue into the 1960s that greenhouse-effect warming of the planet was underway.

While Callendar's claims were largely met with skepticism, he managed to draw attention to the possibility of global warming.

Keeling Curve

Most famous among those research projects was a monitoring station established in 1958 by the Scripps Institution of Oceanography on top of Hawaii's Mauna Loa Observatory.

Scripps geochemist Charles Keeling was instrumental in outlining a way to record CO_2 levels and in securing funding for the observatory, which was positioned in the center of the Pacific Ocean.

The dawn of advanced computer modeling in the 1960s began to predict possible outcomes of the rise in CO_2 levels made evident by the Keeling Curve. Computer models consistently showed that a doubling of CO_2 could produce warming of 2 degrees C or 3.6 degrees F within the next century. nStill, the models were preliminary and a century seemed a very long time away.

When did scientists first warn humanity about climate change?

Climate change warnings are coming thick and fast from scientists; thousands have signed a paper stating that ignoring climate change would yield "untold suffering" for humanity, and more than 99% of scientific papers agree that humans are the cause. But climate change was not always on everyone's radar. So, when did humans first become aware of climate change and the dangers it poses?

Scientists began to worry about climate change toward the end of the 1950s. The scientific community began to unite for action on climate change in the 1980s, and the warnings have only escalated since. However, these recent warnings are just the tip of the melting iceberg; people's interest in how our activities affect the climate dates back thousands of years.

As far back as ancient Greece (1200 B.C. to A.D. 323), people debated whether draining swamps or cutting down forests might bring more or less rainfall to the region, according to Weart's Discovery of Global Warming website, which is hosted by the American Institute of Physics.

The ancient Greek debates were among the first documented climate change discussions, but they focused only on local regions. It was not until a few millennia later, in 1896, that Swedish scientist Svante Arrhenius (1859-1927) became the first person to imagine that humanity could change the climate on a global scale, according to Weart. That's when Arrhenius published calculations in The London, Edinburgh, and Dublin Philosophical Magazine and Journal of Science showing that adding carbon dioxide to the atmosphere could warm the planet.

This work built on the research of other 19th-century scientists, such as Joseph Fourier (1768-1830), who hypothesized that Earth would be far cooler without an atmosphere, and John Tyndall (1820-1893) and Eunice Newton Foote (1819-1888), who separately demonstrated that carbon dioxide and water vapor trapped heat and suggested that an atmosphere could do the same.

Arrhenius' climate change predictions were largely spot on. Human activities release carbon dioxide, methane, and other greenhouse gases that trap radiation from the sun and hold them in the atmosphere to increase temperature like a warming greenhouse, hence the term "greenhouse effect."

However, Arrhenius' work was not widely read or accepted at the time, nor was it even intended to serve as a warning to humanity; it can be viewed as such only in hindsight. At the time, his work simply recognized the possibility of humans influencing the global climate, and for a long time, people viewed warming as beneficial, according to Weart.

There was some coverage of fossil fuels affecting climate in the general media, according to a now-viral 1912 article first published in the magazine Popular Mechanics, USA Today reported. The article, which ran in a few newspapers in New Zealand and Australia later that year, recognized burning coal and releasing carbon dioxide could increase Earth's temperature, noting that "the effect may be considerable in a few centuries."

Why the 1950s?

The scientific opinion on climate change would not begin to shift until two significant experiments some 60 years after Arrhenius' realization. The first, led by scientist Roger Revelle (1909-1991) in 1957 and published in the journal Tellus, found that the ocean will not absorb all of the carbon dioxide released in humanity's industrial fuel emissions and that carbon dioxide levels in the atmosphere could, therefore, rise significantly.

Three years later, Charles Keeling (1928-2005) published a separate study in Tellus that detected an annual rise in carbon dioxide levels in Earth's atmosphere. With carbon dioxide levels known to affect the climate, scientists began to raise concerns about the impact human-related emissions could have on the world.

From there, more studies began highlighting climate change as a potential threat to species and ecosystems around the world. Scientists first began in 1988 to insist that real action should be taken. This occurred at the Toronto Conference on the Changing Atmosphere, where scientists and politicians from around the world gathered to address what was framed as a global threat to Earth's atmosphere, with calls to reduce emissions and knock-on effects such as acid rain.

By the 1990s, most scientists thought the action was necessary, but opposition from fossil fuel companies and ideologists opposed to any government action were effective in obscuring the facts and blocking action. Plus, normal human inertia and unwillingness to do anything without immediate benefits for oneself.

Humanity faces 'grave and mounting threat' of climate change — unless we act, IPCC report reveals more than 270 climate experts authored the UN's IPCC assessment.

Waves crashed over the Newhaven Harbor wall in Newhaven, southern England on February. 18, as Storm Eunice brought high winds across the country. Powerful storms such as this are becoming more frequent due to human-induced climate change.

From food insecurity to our physical and mental health, the impact of climate change is affecting people around the world, and the window is rapidly closing for us to prevent catastrophic and irreversible consequences, according to a new report by the Intergovernmental Panel on Climate Change (IPCC), which evaluates climate science for the United Nations.

Overall, an estimated 3.3 billion to 3.6 billion people inhabit regions that are considered "highly vulnerable to climate change," according to the report. However, the impacts of global warming are unequally distributed, and those who are most vulnerable to climate change are often cut off from resources that could help them to adapt or mitigate risk.

"Today's IPCC report is an atlas of human suffering and a damning indictment of failed climate leadership," António Manuel de Oliveira Guterres, Secretary-General of the United Nations, said at the briefing.

Evidence in the report from more than 34,000 scientific sources shows how extreme storms, droughts, floods, heatwaves, and wildfires — all of which have been increasing in severity and frequency due to climate change — are disrupting food production, interfering with fishing and aquaculture; causing costly damage to cities and infrastructure; and eroding human health.

What's more, that disruption will only worsen the longer we put off taking necessary steps to limit warming to 2.7 degrees Fahrenheit (1.5 degrees Celsius) and help the hardest-hit parts of the world adapt to change that has already happened. This report is a dire warning about the consequences of inaction. It shows that climate change is a grave and mounting threat to our well-being and a healthy planet.

Limiting warming to 2.7 F, would require slashing greenhouse gas emissions globally by 40% and achieving net zero emissions by 2050; instead, the world is on track for emissions to rise an estimated 14% over the coming decade, Guterres said at the briefing.

According to the report, food and water insecurity are on the rise and are affecting millions of people globally, "especially in Africa, Asia, Central, and South America, on small islands and in the Arctic," caused by cascading impacts from weather extremes caused by climate change, such as heat, drought and floods.

On average, global agricultural growth has slowed over the past 50 years as Earth warms, with most of the negative impacts occurring in midlatitude and low latitude regions.

With extreme heat events increasing around the world, there are more annual deaths from heatwaves and respiratory complications linked to already-elevated air pollution.

Climate-related, food-borne and water-borne diseases spread more widely and more rapidly, as do vector-borne illnesses and zoonotic diseases driven by range expansion for the organisms that carry harmful pathogens, according to the report.

Data from North America shows that climate change harms mental health, too. People who have lost their homes, livelihoods, or loved ones in floods and wildfires may be affected by post-traumatic stress disorder, while other impacts of climate change, such as food insecurity, can likewise affect mental wellbeing.

Watching news stories or reading about the damage caused by climate change — and worrying about what is to come — can also negatively impact mental health,

even when the person following the news has not experienced destructive climate change firsthand.

1970s Scare: A Cooling Earth

In the early 1970s, a different kind of climate worry took hold: global cooling. As more people became concerned about pollutants, people were emitting into the atmosphere, some scientists theorized that pollution could block sunlight and cool Earth.

Earth did cool somewhat between 1940-1970 due to a postwar boom in aerosol pollutants which reflected sunlight away from the planet. The idea that sunlight-blocking pollutants could chill Earth caught on in the media, as in a 1974 Time magazine article titled "Another Ice Age?"

But as the brief cooling period ended and temperatures resumed their upward climb, warnings by a minority of scientists that the Earth was cooling were dropped. Part of the reasoning was that while smog could remain suspended in the air for weeks, CO_2 could persist in the atmosphere for centuries.

Global Warming Gets Real in 1988

The early 1980s would mark a sharp increase in global temperatures. Many experts point to 1988 as a critical turning point when watershed events placed global warming in the spotlight.

The summer of 1988 was the hottest on record, although many since then have been hotter. 1988 also saw widespread drought and wildfires within the United States.

Scientists sounding the alarm about climate change began to see media and the public paying closer attention. NASA scientist James Hansen delivered testimony and presented models to congress in June of 1988, saying he was "99 percent sure" that global warming was upon us.

IPCC

One year later, in 1989, the Intergovernmental Panel on Climate Change (IPCC) was established by the United

Nations to provide a scientific view of climate change and its political and economic impacts.

As global warming gained currency as a real phenomenon, researchers dug into possible ramifications of a warming climate. Among the predictions were warnings of severe heat waves, droughts, and more powerful hurricanes fueled by rising sea surface temperatures.

Other studies predicted that as massive glaciers at the poles melt, sea levels could rise between 11 and 38 inches (28 to 98 centimeters) by 2100, enough to swamp many of the cities along the east coast of the United States.

Kyoto Protocol: United States In, Then Out

Government leaders began discussions to try and stem the outflow of greenhouse gas emissions to prevent the direst predicted outcomes. The first global agreement to reduce greenhouse gases, the Kyoto Protocol, was adopted in 1997.

The protocol, which was signed by President Bill Clinton, called for reducing the emission of six greenhouse gases in 41 countries plus the European Union to 5.2 percent below 1990 levels during the target period of 2008 to 2012.

In March 2001, shortly after taking office, President George W. Bush announced the United States would not implement the Kyoto Protocol, saying the protocol was "fatally flawed in fundamental ways" and citing concerns that the deal would hurt the U.S. economy.

An Inconvenient Truth

That same year, the IPCC issued its third report on climate change, saying that global warming, unprecedented since the end of the last ice age, is "very likely," with highly damaging future impacts. Five years later, in 2006, former Vice President and presidential candidate Al Gore weighed in on the dangers of global warming with the debut of his film An Inconvenient Truth. Gore won the 2007 Nobel Peace Prize for his work on behalf of climate change.

Politicization over climate change, however, would continue, with some skeptics arguing that predictions presented by the IPCC and publicized in media like Gore's film were overblown.

Among those expressing skepticism over global warming was future U.S. president Donald Trump. On November 6, 2012, Trump tweeted "The concept of global warming was created by and for the Chinese to make U.S. manufacturing non-competitive."

Paris Climate Agreement: the United States In, Then Out

The United States, under President Barack Obama, signed another milestone treaty on climate change, the Paris Climate Agreement, in 2015. In that agreement, 197 countries pledged to set targets for their own greenhouse gas cuts and to report their progress.

The backbone of the Paris Climate Agreement was a declaration to prevent a global temperature rise of 2 degrees C (3.6 degrees F). Many experts considered 2 degrees C of warming to be a critical limit, which, if surpassed will lead to an increased risk of more deadly heat waves, droughts, storms and rising global sea levels.

The election of Donald Trump in 2016 led to the United States declaring it would withdraw from the Paris treaty. President Trump, citing the "onerous restrictions" imposed by the accord, stated that he could not "in good conscience support a deal that punishes the United States."

That same year, independent analyses by NASA and the National Oceanic and Atmospheric Administration (NOAA) found Earth's 2016 surface temperatures to be the warmest since modern record-keeping began in 1880.

And in October 2018, the U.N.'s Intergovernmental Panel on Climate Change issued a report that concluded "rapid, far-reaching" actions are needed to cap global warming at 1.5 Celsius (2.7 Fahrenheit) and avert the direst, irreversible consequences for the planet.

Greta Thunberg and Climate Strikes

In August 2018, Swedish teenager and climate activist Greta Thunberg began protesting in front of the Swedish Parliament with a sign: "School Strike for Climate." Her protest to raise awareness of for global warming caught the world by storm and by November 2018, over 17,000 students in 24 countries were participating in climate strikes.

By March 2019, Thunberg was nominated for a Nobel Peace Prize. She participated in the United Nations Climate Summit in New York City in August of 2019, famously taking a boat across the Atlantic instead of flying to reduce her carbon footprint.

The UN Climate Action Summit reinforced d that "1.5°C is the socially, economically, politically and scientifically safe limit to global warming by the end of this century," and set a deadline for achieving net zero emissions by 2050.

CHAPTER TWO
CLIMATE CHANGE AND OUR HEALTH

What is climate change, and how is it affecting Earth?

Climate change is any long-term alteration in average weather patterns, either globally or regionally. Climate change has occurred many times in Earth's history, and for many different reasons. The changes in global temperature and weather patterns are seen today, however, are caused by human activity. And they are happening much faster than the natural climate variations of the past.

Scientists have many ways to track climate over time, all of which make it clear that today's climate change is linked to the emission of greenhouse gases, such as carbon dioxide and methane. These gases trap heat from the sun's rays near Earth's surface, much like the glass walls of a greenhouse keep heat inside. Small changes in the proportions of greenhouse gases in the air can add up to major changes on a global scale.

On average, the effect of greenhouse gases is to increase global temperatures. This is why climate change is sometimes called global warming. However, most researchers today prefer the term climate change because of the variability of weather and climate across the globe. For example, warming global average temperatures might alter the flow of the jet stream, the major air current affecting North American weather, which could, in turn, lead to seasonal periods of extreme cold in some areas.

"It's important for people to realize that there is a lot of variability from place to place on the Earth in terms of the temperature," said Ellen Mosley-Thompson, a paleoclimatologist at the Byrd Polar and Climate Research Center of The Ohio State University. "When we talk about global climate change, we are talking about temperature changes over large areas."

How do scientists know climate change is real?

The effects of global warming are visible. The climate of the past is recorded in ice, sediments, cave formations, coral reefs, and even tree rings.

Researchers can look at chemical signals — such as the carbon dioxide trapped in bubbles inside glacial ice — to determine what atmospheric conditions were like in the past.

They can study microscopic fossilized pollen to learn what vegetation used to thrive in any given area, which in turn can indicate what the climate was like. Scientists can also measure tree rings to get a season-by-season record of temperature and moisture. Ratios of chemical variants of oxygen in corals and stalactites and stalagmites can reveal past precipitation patterns.

Different types of natural records can reveal different clues about the climate of the past. Ocean sediments do not carry season-by-season or even year-by-year levels of detail, but they can provide blurrier pictures of climate dating back millions of years.

Tree records are relatively short but incredibly detailed. And ice can be chock-full of information: Not only do glaciers capture atmospheric gases in the form of air bubbles, but they also trap dust and other sediments, pollen grain, volcanic ash, and more. As the ice gets older and more compressed, the record can become fuzzy, but newer ice can provide a year-by-year look at what the climate was doing.

The most recent changes in the climate — since the beginning of the Industrial Revolution — have also been tracked directly. Record-keeping of things like land temperature began to improve in the late 1800s, and ship captains began to keep a wealth of ocean-based weather data in their logs.

The advent of satellite technology in the 1970s provided an explosion of data, covering everything from ice extent at the poles to sea surface temperature to cloud coverage.

Overall, the Earth is warming up because of human-caused climate change. But climate change also causes seasonal periods of extreme cold.

How is the climate changing?

Taken together, these records show that the modern climate is undergoing a swift departure from the patterns of the past.

Before the Industrial Revolution, there were about 280 carbon dioxide molecules for every million molecules in the atmosphere, or 280 parts per million (ppm). As of 2021, the global average level of CO_2 was 419 ppm — more than 100 ppm higher than the level has been in the last 800,000 years, and up 6.5 ppm from 2020, according to the National Oceanic and Atmospheric Administration (NOAA). The last time atmospheric carbon reached today's levels was 3 million years ago, according to NOAA.

The rate of change in today's atmospheric carbon is also faster than in the past, according to NOAA. The rate of increase was 100 times faster over the past 60 years than any time in the last million years or so — a period that saw eight major climate flip-flops between glacial cycles, in which ice expanded from the poles into the middle latitudes, and interglacial cycles, in which the ice retreated to where it is today.

And the rate continues to increase. In the 1960s, atmospheric carbon went up by an average of 0.6 ppm a year. In the 2010s, it rose by an average of 2.3 ppm per year.

The heat-trapping ability of all that extra carbon has translated to rising global average temperatures. According to the National Aeronaautics and Space Administration's (NASA) Goddard Institute for Space Studies (GISS), Earth's average temperature has risen by just over 2 degrees Fahrenheit (1 degree Celsius) since 1880, a measurement accurate to within a tenth of a degree Fahrenheit.

As with the rate of atmospheric carbon increase, the rate of global temperature increase is also speeding up, according to NASA's Earth Observatory: Two-thirds of the warming that's taken place since 1880 has occurred since 1975.

What are the impacts of climate change?

This warming has caused changes in Earth's ecosystems and environments. Some of the most dramatic changes have occurred in the Arctic, where sea ice is on the decline.

Record-breaking ice extent lows have been the new normal since 2002, according to NASA, and studies are finding that even the oldest, multiyear sea ice is thinning rapidly.

Summer 2020 was one of the worst years ever for summer sea ice extent, with only one year on record – 2012 – exhibiting a lower ice extent. Scientists now expect the first ice-free Arctic summer sometime between 2040 and 2060.

Glaciers are retreating globally, particularly in the middle latitudes, Mosley-Thompson said. Montana's Glacier National Park was home to 150 glaciers in 1850. Today, there are only 25. Mosley-Thompson and her team estimate that the last tropical glaciers will disappear within the next decade.

Melting ice and the expansion of ocean waters due to heat has already contributed to rising sea levels. According to NOAA, the global average sea level has risen 8 to 9 inches (21 to 24 centimeters) since 1880.

The rate of rising is increasing, from 0.06 inches (1.4 millimeters) per year in the 20th century to 0.14 inches (3.6 mm) per year from 2006 to 2015. According to NOAA, this sea level rise has translated to a 300% to 900% increase in high-tide flooding in coastal areas of the United States. Ocean water absorbs carbon dioxide from the atmosphere, which creates a chemical reaction that causes ocean acidification.

The global average pH of ocean surface waters has decreased by 0.11 since the Industrial Revolution began — a 30% increase in acidity — according to NOAA's Pacific Marine Environmental Laboratory. Increasing ocean acidity makes it more difficult for corals to build their carbonate skeletons and for shelled animals such as clams and some types of plankton to survive.

Climate change is even affecting the timing of spring-like weather. The earliest spring (as defined by plant growth and temperatures) on record in the United States was in March 2012. Climate models now suggest that such early springs could be the norm by 2050.

But late freezes will likely still occur, creating conditions in which plants could leaf out early and then be damaged by cold temperatures. Climate models also predict the exacerbation of alarming trends in droughts and wildfires due to warmer temperatures.

Models are a key tool for climate scientists, said Kathie Dello, a state climatologist for North Carolina. There is no way to compare different futures for Earth in the real world, Dello said, but models enable scientists to create virtual versions of the planet to test different scenarios.

Though the Earth system is complicated, these computer models have proved capable of predicting future climate trends. A 2020 paper in the journal Geophysical Research Letters found that climate model predictions published between the 1970s and 2010 were accurate when compared with the actual warming that occurred after publication.

Can we reverse climate change?

A growing number of business leaders, government officials, and private citizens are concerned about climate change and its implications, and they are proposing steps to halt and reverse the trend.

While some argue that 'the Earth will heal itself,' the natural processes for removing this human-caused CO_2 from the atmosphere work on the timescale of hundreds of thousands to millions of years. So, yes, the Earth will heal itself, but not in time for our cultural institutions to be preserved as they are. Therefore, in our self-interests, we must act in one way or another to deal with the climate changes we are causing."

If all human greenhouse gas emissions stopped immediately, Earth would likely still experience more warming, some studies suggest, because carbon dioxide remains in the atmosphere for hundreds of years.

Some proposals could theoretically reverse some of this "locked in" warming by removing carbon dioxide from the atmosphere, such as carbon capture and storage, which involves injecting carbon into underground reservoirs. Advocates argue that carbon capture and storage is technologically feasible, but market forces have prevented widespread adoption.

Whether or not removing already-emitted carbon from the atmosphere is feasible, preventing future warming requires humans to stop causing the emission of greenhouse gases. The most ambitious effort to forestall warming thus far is the Paris Agreement.

This nonbinding international treaty, which came into effect in November 2016, aims to keep warming "well below 2 degrees Celsius (3.6^{0}F) above pre-industrial levels and to pursue efforts to limit the temperature increase even further to 1.5 degrees Celsius (2.7^{0}F)," according to the United Nations.

Each signatory to the treaty agreed to set their own voluntary emissions limits and to make them stricter over time. Climate scientists said that the emissions limits that were outlined in the agreement would not keep warming as low as 1.5 or even 2 degrees C, but that it would be an improvement over the "business-as-usual" scenario in which no changes are made to cut greenhouse gas emissions.

Under the Obama administration, the United States pledged to limit greenhouse emissions to less than 28% of 2005 levels by 2025. However, President Donald Trump announced soon after his election that his administration would not honor the Paris Agreement. The Trump administration began the formal withdrawal process from the agreement in 2019. Upon assuming the presidency in 2021, Joe Biden re-committed the U.S. to the Paris Agreement.

A 2021 study found that greenhouse emissions have already "locked in" enough warming such that Earth will warm more than 3.6^0F, pushing past the Paris Agreement's goal. However, curbing emissions could still slow the temperature rise to a more manageable rate and reduce the ultimate peak.

Several state and local governments have launched their efforts to combat climate change. For instance, 24 states and Puerto Rico have joined the U.S. Climate Alliance, pledging to meet goals set in the Paris Agreement regardless of politics at the federal level. The federal government, even when it is operating well, is not the nimblest institution. But states and cities are a little more flexible.

8 ominous climate milestones reached in 2021

Signs of accelerating global warming abounded this year. The disastrous consequences of burning fossil fuels and pumping greenhouse gases into Earth's atmosphere are everywhere around us. Study after study directly links human-caused climate change to more powerful and wetter storms, longer and more intense droughts, and rising sea levels that threaten coastal communities worldwide. *And 2021 made the accelerating pace of climate change painfully clear.*

While we still have time to mitigate the worst climate change impacts, that can happen only if we drastically and quickly reduce greenhouse gas emissions — and soon.

Here are eight signs in 2021 that the window to avoid climate catastrophe is closing (though it is still not too late to change course).

Paris Agreement warming targets surpassed

When world leaders signed the climate action pledge known as the Paris Agreement in 2015, they committed to long-term and short-term plans for reducing the consumption of fossil fuels and the production of greenhouse gasses linked to climate change.

Their goal: reduce global warming to 3.6 degrees Fahrenheit (2 degrees Celsius). But global average temperatures have already climbed to about 1.8^0F (1^0C) warmer than they were during pre-industrial times, and the 2015 goal is already out of reach.

And the warmer Earth gets, the more warming accelerates; as the planet loses ice and snow, it reflects less heat trict into space and absorbs it instead, scientists reported in January in the journal Nature Climate Change.

Record-breaking heat in 2020

At the start of 2021, NASA climate scientists announced that 2020 ranked alongside 2016 as the hottest year of all time. Researchers at NASA's Goddard Institute for Space Studies (GISS) in New York stated in January that 2020's global average surface temperatures were warmer than the 20th-century average by 1.84 F (1.02^0C).

However, in a separate assessment, researchers with the National Oceanic and Atmospheric Administration (NOAA) reported that 2020 was the second-hottest year after 2016, with temperatures that were 1.76^0F (0.98^0C) higher than average — just 0.04 F (0.02^0C) cooler than 2016's average temperatures.

Though the conclusions of the two agencies presented slight variations, both concurred that the current warming trend on Earth is unprecedented, with average global temperatures on the rise for more than 50 years.

Faster sea level rise

We have likely been underestimating how quickly sea level rise could happen, a February study showed. Prior models estimated that by the year 2100, the global sea-level average would likely rise by 3.61 feet (1.10 meters), but scientists now suggest that oceans will rise even more rapidly than that, based on sea level rise events in Earth's distant past.

By evaluating historical data and looking at how quickly seas rose and fell as ancient Earth warmed and cooled, researchers could then estimate a rate for the future sea-level rise that was unexplored in previous computations. The scientists found that existing sea-level models predicted more conservative maximums than the new models did.

Gulf Stream slowdown

Earth's climate is regulated by ocean currents and one of the most important of these is the Gulf Stream, which acts like a giant conveyer belt transporting heat around the ocean.

However, due to human-induced climate change, the Gulf Stream has slowed dramatically and could stop completely by 2100, if global warming continues at its current pace, new research found.

The Gulf Stream regulates climate and weather by circulating warm, salty water around the planet. But as Earth warms, melting fresh water ice pours into the ocean, lowering the salinity of the water and disrupting the current's flow. Should the Gulf Stream falter and fail, it could trigger more extreme weather, such as cyclones and heat waves, and may accelerate sea level rise in coastal Europe and North America.

Human influence 'unequivocal'

The evidence that humans are driving climate change is crystal clear, according to a report authored by over 200 climate experts who reviewed more than 14,000 studies. In August, the United Nations Intergovernmental Panel on Climate Change (IPCC), the UN body focusing on climate science, released the first installment of the IPCC's Sixth Assessment Report, which stated that human-driven changes are affecting all of Earth's planetary systems in ways that are "widespread and rapid."

Hundreds of researchers co-authored the report, finding that the burning of fossil fuels has pumped so much CO_2 into the atmosphere that global warming is advancing at a rate that is unprecedented in the past 2,000 years.

Carbon factory rainforests

Wildfires in the Amazon are polluting the air with greenhouse gases faster than the surviving trees can absorb them. Tropical rainforests are often called the "lungs of the planet" because they produce oxygen and absorb atmospheric carbon dioxide (CO_2).

But that pattern has been reversed; the Amazon rainforest is now emitting more CO_2 than it absorbs, releasing more than 1.1 billion tons (1 billion metric tons) of CO_2 per year, while absorbing only about half a billion tons. Large-scale human disturbances were responsible for the shift, with wildfires producing much of the excess CO_2 — and most of the wildfires were deliberately set to clear land for industry and agriculture.

'Last Ice Area' melting away

This photo of sea ice on the Wandel Sea north of Greenland was taken August 16, 2020, from the German icebreaker Polarstern, which passed through the area as part of the year-long MOSAiC Expedition. This area used to remain fully covered in ice throughout the year.

To the north of Greenland lies a frozen zone that previous research suggested would remain mostly frozen even as

Earth's climate warmed. But even this so-called Last Ice Area may not survive the current rate of global warming.

In 2020, ice cover in the Wandel Sea in the eastern part of the Last Ice Area reached its lowest since record-keeping began, with about 50% of the sea ice melting away during the summer months. When scientists recently analyzed the ice loss, they discovered that year-round melt caused by rising global temperatures was reducing the overall thickness of the region's permanent ice over time.

This means that prior models predicting the Last Ice Area's longevity have likely been underestimating the rate of ice loss — and the area could become ice-free as soon as 2040.

Earthshine gets darker

Scientists recently investigated a previously unexamined consequence of climate change: a decrease in Earth's brightness. Our planet reflects sunlight onto the surface of the moon's dark side, in a phenomenon known as "earthshine."

Using satellite views, researchers measured earthshine and tracked variations in brightness based on the reflectiveness of clouds in the atmosphere, and of water, land and snow, and ice cover on Earth's surface. They then compared datasets of earthshine observations with other datasets that recorded changes in Earth's cloud cover.

The researchers saw that over the past two decades, Earth's light has dimmed by approximately 0.5% — it now reflects about half a watt less light per square meter. The scientists also found that the dimming corresponded with a decline in bright low-altitude clouds over the eastern Pacific Ocean.

Clouds are a complicated piece of the climate puzzle, but this drop is likely linked to other atmospheric changes caused by climate change, the scientists reported in August in the journal Geophysical Research Letters.

Humanity faces a' grave and mounting threat' of climate change

We have more evidence now of that dependency, and that a lot of the adaptation options involve some level of restoring and protecting natural ecosystems, and developing society in ways that are more coupled into a matrix of a natural landscape. In February, wildfires fueled by severe drought consumed forests, grasslands, and wetlands in northeastern Argentina, burning an estimated 40% of the Ibera National Park.

However, many natural ecosystems are already nearing collapse due to stresses from global warming, and mounting evidence shows that our adaptation options will decline sharply as natural systems fail. Earth has already warmed to nearly 2.0^0F (1.09^0C) above pre-industrial average temperatures, and the impact on diverse ecosystems is far more negative and widespread than prior reports anticipated.

Some of the changes outlined in the new report were unexpected at 2.0^0F of warmings, such as diseases emerging in North American forests, the first extinctions of species due to climate change, and mass mortality events in trees and mammals due to heatwaves and drought. With increased insect pest outbreaks, more tree deaths and wildfires, and the loss of permafrost and the drying of peatlands, Earth's biosphere is becoming less capable of absorbing greenhouse gases that are emitted by humans.

Regions that were once reliable carbon sinks — absorbing atmospheric carbon dioxide (CO_2) — such as old-growth Amazon rainforests and permafrost expanses in undisturbed areas of North America and Siberia, are in some areas transforming into CO_2 factories that produce more carbon than they absorb.

And as these changes are already underway with present warming levels, reversing these processes will likely be more difficult than models predicted should warming climb past the target of 2.7^0F. Because adapting to a warming world — and capping warming at 2.7^0F — will require global cooperation and significant investments from the world's nations, the challenge can seem overwhelming on a personal level.

However, even seemingly small actions can help to shape change across communities and will help us adapt as Earth warms. There are lots of actions individuals can take separate from governments — checking on older adults and other vulnerable populations during heatwaves is one of many examples. Further, there are thousands of NGOs (nongovernmental organizations) across the United States, many of which are working on issues related to adaptation and sustainability, including vulnerability reduction.

Engaging with local NGOs on topics of interest is an excellent opportunity for moving adaptation forward.

Indeed, the IPCC report indicates that a whole of society response— one that includes individuals, communities and governments — will be essential if we are to succeed in reducing fossil fuel reliance, limiting global warming and adapting to climate change challenges.

We all need to opt into the solution. How we use our sense of agency in the world, how we engage with governance processes, how we engage with leadership in our communities, the kind of priorities we express about the kind of world we want to see, which will influence policies — all of this is critical. The individual can play a vital role.

Human Health Impacts of Climate Change

Climate change impacts human health in both direct and indirect ways. Extreme heat waves, rising sea level, changes in precipitation resulting in flooding and droughts, and intense hurricanes can directly cause injury, illness, and even death. The effects of climate change can also indirectly affect health through alterations to the environment. For example, worsening air pollution levels can have negative impacts on respiratory and cardiovascular conditions.

Changes in temperature and rainfall can alter the survival, distribution, and behavior of insects and other species that can lead to changes in infectious diseases. Increases in precipitation, storm surge, and sea temperature can lead to more water-related illnesses.

Climate change can also affect food safety, exposing people to contaminated foods that can result in foodborne illnesses. In addition, climate change can affect mental health and well-being.

Exposure to climate-related hazards can include biological, chemical, or physical stressors and can differ in time, locations, populations, and severity. These are referred to as exposure pathways. These threats can occur simultaneously, resulting in compounding health impacts. Climate change threats may also accumulate over time, leading to longer-term changes in resilience and health.

Climate change can affect human health by changing the severity, duration, or frequency of health problems and by creating unprecedented or unanticipated health problems or health threats in places or populations where they have not previously occurred.

While everyone is exposed to climate-related health threats, not everyone experiences the same harms. Individuals may experience greater risk from climate-related health effects because: they have greater exposure to climate-related hazards; they are more sensitive to the effects of climate stressors; their own present state of health and wellbeing; or they do not have sufficient capacity or resources to cope or remove themselves from harm. An effective public health response to mitigate the risks of climate change is essential to preventing injuries and illnesses and enhancing overall public health preparedness.

Still not too late

While we can not turn back the clock and reset Earth's climate to conditions that predate the Industrial Age, that does not mean there is nothing we can do about climate change. Under the current warming trend, by the year

2050 Earth will become more than 3.6^0F (2^0C) hotter on average.

However, if we reduce fossil fuel use and limit the rise of global temperature averages to no more than 2.7^0F (1.5^0C) above pre-Industrial levels, we can still slow or stop some of the global changes that are already underway, such as sea level rise and extreme weather events, according to the IPCC report.

If current warming continues, sea level rise could reach 7 feet (2 meters) by 2100. But reducing greenhouse gases and allowing Earth to cool down could slow that process by thousands of years, climate experts wrote in the report. Scientists are also working to develop new computer models to create updated predictions about timescales for ice melt and sea level rise, and to explore how human communities — especially the most vulnerable ones — might adapt to these changes.

But in order to get there, humanity needs to take action, and that begins with dramatically curbing our use of fossil fuels on a global scale, and enacting legislation to rebuild infrastructures around sustainable energy sources, Michael Mann, a climatologist at The Pennsylvania State University, explained.

Can we adapt?

The good news is that humans are an adaptable species, and people can adjust to life in a warming world; in fact, growing public and political awareness of climate impacts and risks has resulted in at least 170 countries and many cities including adaptation in their climate policies and planning processes. But those strategies can vary widely depending on location, and may be greatly constrained by inequity and poverty.

One of the key findings is that many viable adaptation options rely on natural ecosystems, such as wetlands and inland rivers that help mitigate flooding from rising sea levels in coastal areas.

Climate-induced Natural Disaster

While there may not be an exact blueprint for what climate change is going to do to each of our lives, experts have some solid guesses that, combined with some good old common sense, can help each of us prepare for our new normal.

The evidence is clear. Climate change is making natural disasters more frequent, more severe, and more expensive. We are getting freak heat waves and freak snowstorms, devastating droughts and historic downpours, flooding and water shortages. Everything is changing simultaneously: oceans, atmosphere, plants, animals, permafrost, weather, seasons, insects, people.

Because your risk of natural disaster is completely dependent on where you live, what is most important is that you understand what disasters you, personally, may face, and do not just rely on what disasters you have faced in the past—that is not an accurate assessment anymore. You can do this by researching your city or county's emergency preparedness tips and making sure you understand the basics of surviving an earthquake, tornado, hurricane, flood, or wildfire.

No matter where you live, you should make sure your homeowner or renter insurance covers the disasters you are at risk for. He also points out that you don't need to live on a coast to be at risk for flooding, and homeowners' insurance does not cover flooding. After your insurance is squared away, undergo a prepping for two weeks of having no water, food, or power, packing a "go bag" to sustain you for a couple of days outside of your home, and making a plan with your family about where to meet if cell towers are not working. There is straight forward book (The Beginner's Big Book With Over 40 Healthy Make-Ahead Recipes) on prepping by Kathleen Colquitt, you will find it very useful.

The last piece of advice is the simplest: get the Red Cross Emergency mobile application downloaded. It is free and will give you early warning about disasters. The most tragic way to die in a fire, flood, or hurricane is in your home because you never got the word to evacuate.

Supply Chain Breakdown and Food Shortages

Whether or not you agree with experts who say that climate change could bring about a Roman Empire–esque societal collapse, it is clear that shortages and supply chain disruptions are on the increasingly warm horizon.

As Covid-19 showed us, those disruptions can impact anything from medical supplies to car parts to finding a winter coat. But the most concerning shortages that we face are access to food and water. A 2019 UN report warns of a looming food crisis, and drought already threatens 40 percent of the world's population, according to the WHO, and over 80 million people in the United States, according to the US government's Drought Information system.

It is also suggested that climate change will cause rising food prices, greater food insecurity, and may lead to micronutrient deficiencies in more people. While there may be little you can do to impact the global food chain, you can start in your own backyard by planting a fruit tree or starting a garden, learning how to grow climate-appropriate vegetables, and making sure your pantry is fully stocked with two weeks of water and food, along with any necessary medical supplies. It is also important to assume you will not have warning before a food and water shortage, so do not put off stocking up until it is too late.

Becoming Resilient Together

Resilience may be an overused term when we talk about climate change, but for most of us, it is grossly lacking in how prepared we are to care for ourselves, our loved ones, and our property if emergency workers are not able to assist us. Barely half of Americans can perform CPR (Cardiopulmonary resuscitation), only 17 percent know how to build a fire, and just 14 percent feel confident in their ability to identify edible plants and berries.

Basic skills like learning how to operate a two-way radio, knowing the smartest escape route out of your city or neighborhood, or being able to change a bike tire may sound simple, but can be the difference between life and death in a disaster.

Perhaps the most effective way to take care of yourself is to get close to others. According to FEMA, 46 percent of people expect to rely a great deal on people in their neighborhood for assistance within the first 72 hours after a disaster. Prepping is not a lone wolf activity. It is important that your immediate neighbors know your name and who is in your family, including pets so they can inform first responders in the case of an earthquake or a fire.

In the event of supply chain disruptions, your neighbors may be your only access to vital supplies like batteries or extra diapers. Building connections in your local community is also a great way to build an informal service network, because who knows when you may need help with an injury or a home repair. Always remember that community wins in 99 percent of situations.

Expert Recommendation About Climate Crisis

The global strike on Friday September 20, 2019 was the largest demonstration for climate action in history. The movement that started with Swedish teenager Greta

Thunberg in August 2018 has now mobilised millions, while Extinction Rebellion and other protest groups have escalated their campaigns on streets around the world. From the efforts of activists in different countries, radical "Green New Deals" are emerging as a bold, political response to the climate crisis.

"The world is waking up," Thunberg told world leaders at the recent UN Climate Action Summit. "And change is coming whether you like it or not."

But how long that awakening takes could be decisive for warming this century. A report published ahead of the summit declared that the impacts of climate change are accelerating – growth rates of carbon dioxide (CO_2) in the atmosphere were nearly 20% higher in the 2015-19 period than the previous five years, while the average rate of sea level rise has increased to 5mm per year over roughly the same period.

The debate over whether climate change is happening is over, and the conversation about what should be done is beginning in earnest.

Nowhere are the effects of climate change more visible than in the Arctic – a region that's estimated to be warming at least twice as fast as the global average. Arctic sea ice reached its second-lowest extent on record in September 2019 – a mere 1.6m square miles.

A new report by the Intergovernmental Panel on Climate Change (IPCC) has revealed the changes that are underway in the oceans and the ice-covered regions of the world, and the message is stark.

According to Cassandra Brooks, University of Colorado Boulder, Glaciers and ice sheets are shrinking. Global sea level is rising at more than twice the rate of the 20th century. The ocean is warming, becoming more acidic and losing oxygen. Fifty percent of coastal wetlands have been lost over the past 100 years. Species are shifting, biodiversity is declining and ecosystems are losing their integrity and function.

Earth's oceans have absorbed more than 90% of the excess heat in the global climate system, but not without repercussions. Marine heatwaves are causing coral reefs to bleach and mass die-offs of fish and other wildlife. Not only does this added ocean heat wreak havoc on marine ecosystems, but it is changing the relationship between coastal communities and the ocean.

Wildlife populations are struggling to adapt to the disruption, both in the oceans and on land. As a result, a landmark report from an international team of biodiversity experts earlier in 2019 called for "transformative change" to economies and societies to prevent up to a million species going extinct.

Five steps to restrain emissions right now

Rebecca Willis of Lancaster University explained that the UK, Norway, Sweden and France have written a target of net-zero emissions into law, but targets are, by definition, a statement of nothing more than intent. In and of themselves, they do not remove a single molecule of carbon dioxide from the atmosphere.

Limiting global temperature rise to 1.5°C means the world has to be at net-zero carbon emissions by 2050. But it's cumulative emissions that count – if countries leave decarbonization, it will be too late. Rebecca Willis, a professor of climate and energy policy at Lancaster University in the UK, outlines five measures that can begin to restrain emissions right now:

- Task all government departments with a climate agenda so that they're obligated to show how their policies will contribute to emissions cuts.
- Engage the public in developing climate strategies with citizens' assemblies and involve workers in designing policy to ensure a "just transition" from carbon-intensive industries like coal mining.
- Enact "symbolic policies" to set the investment climate and catalyze radicals catalyze– ban advertising for petrol cars for instance, or allow

people to generate and sell renewable energy at home.
- Keep all remaining fossil fuels in the ground – ban oil and gas exploration and end the US$5.2 trillion spent on subsidizing the subsidizing industry each year.
- Set aside negative emission technologies – carbon capture and storage might play a role in absorbing greenhouse gases from the atmosphere in the future, but they do not exist at a meaningful scale yet. Relying on them distracts resources from cutting emissions now.

CONCLUSION

There are technologies that already exist which are very good at capturing CO_2. Plants have perfected the process of turning carbon in the air into solid, living material over billions of years. Natural climate solutions mobilise this potential by restoring forests, wetlands and other habitats.

But it's not as simple as letting nature do all the hard work. The right tree should be planted at the right time in the right place.

Tree species that grow fast are able to store carbon more rapidly, but slower, bigger trees will ultimately store more in the long run. Non-native trees may be better at absorbing carbon efficiently than native species, so keep an open mind to their role. When choosing a species to plant, trust local foresters and the community that will have to live alongside and nurture them.

Natural climate solutions can also help society adapt to the inevitable changes ahead. The value of rewilding here is two-fold: removing the CO_2 that exacerbates climate change while mitigating its impacts.

Years ago, climate journalists threw away the notion of putting scientific fact and climate denier opinion on equal ground. Other media are finally catching up and providing analysis and verification, and improving the scientific literacy of their audience. They are connecting climate change to local stories, focusing on solutions (instead of "doom and gloom" coverage) and adjusting their language to describe the "climate crisis". That's why "system change not climate change" has become the rallying cry of the young climate strikers.